Tabletop
Gardens

Tabletop
Gardens

How to Make Long-Lasting Arrangements for Every Season

BAYLOR CHAPMAN

PHOTOGRAPHS BY PAIGE GREEN

ARTISAN BOOKS | NEW YORK

Library of Congress Cataloging-in-Publication Data

Names: Chapman, Baylor, author.
Title: Tabletop Gardens / Baylor Chapman.
Description: New York, NY : Artisan, a division of Workman Publishing Co.,
    Inc., [2021] | Includes index.
Identifiers: LCCN 2020042967 | ISBN 9781648290336 (hardcover)
Subjects: LCSH: Flower arrangement in interior decoration. | Plants,
    Potted. | House plants.
Classification: LCC SB449.3.P65 C46 2021 | DDC 635.9/66 –dc23
LC record available at https://lccn.loc.gov/2020042967

Design by Nina Simoneaux

Artisan books are available at special discounts when purchased in bulk for premiums
and sales promotions as well as for fund-raising or educational use. Special editions or
book excerpts also can be created to specification. For details, contact the Special Sales
Director at the address below, or send an e-mail to specialmarkets@workman.com.

For speaking engagements, contact speakersbureau@workman.com.

Published by Artisan
A division of Workman Publishing Co., Inc.
225 Varick Street
New York, NY 10014-4381
artisanbooks.com

Artisan is a registered trademark of Workman Publishing Co., Inc.
Published simultaneously in Canada by Thomas Allen & Son, Limited

Printed in China
First printing, March 2021

10 9 8 7 6 5 4 3 2 1

This book has been adapted from *The Plant Recipe Book* (Artisan, 2014)

To Lila,
for imparting her love
of natural beauty

CONTENTS

Introduction   **11**

Getting Started   **14**

THE GARDENS

Winter Bulbs	**34**
Ground Cover	**37**
A Forest in Miniature	**38**
Pattern Play	**41**
Pretty in Pink	**42**
A Basket of Begonias	**45**
Purple Haze	**48**
An Everlasting Arrangement	**50**
Natural Neons	**53**
English Garden	**56**
Blushing Blooms	**59**
Classical Asymmetry	**60**
Untamed Jungle	**63**
Dancing Ladies	**66**
Tropical Treasure Chest	**69**
Wooded Whimsy	**73**
Rugged Blooms	**74**
A Study in Contrasts	**76**
Lush Lantern	**80**
Ruffles and Roses	**83**
Jewel Tones	**86**

Long and Lean   89

Woodland Violets   90

Fiery Sunset   93

Vintage Trough   94

Springtime in a Bowl   97

All in a Row   98

An Early Summer Centerpiece   101

A Dramatic Duo   102

Velvet Menagerie   105

Acknowledgments   108
Index   110

INTRODUCTION

Blooming centerpieces are my go-to design element for all occasions. Their on-the-spot impact is unbeatable—add them to any room and suddenly it feels polished. It's the thoughtful combination of a plant (or plants) and their vessels that can make them go from "just flowers" to statement pieces—perfect for a welcoming entryway's focal point or a dining room centerpiece for a celebratory occasion. Floral arrangements, when done right, evoke a feeling of luxury, while being both accessible and requiring minimal effort. The intricacy and scent of these beauties may feel lavish (and they are often used to decorate for special occasions), but they also connect us to nature in a way that is downright essential.

And with the projects in this book, you can bring the garden indoors in a carefully designed, meaningful, and long-lasting way. While cut-flower arrangements fade with time, these planted centerpieces use living, growing plants to create the same impact, for longer. The projects you'll see in the following pages will range from incredibly simple and classic to more complicated and ornate. Some designs will last the life of the plants in them (years, in some cases) while others are more temporary—but even those can be disassembled and repotted or planted in a garden.

Most of the plants used in this book are easy to find (African violets, for example), while a few more unique species (coppertone sansevieria) are added in for variety. Some are seasonal, like amaryllis and hyacinth, while a fern or peperomia is easy to find almost anywhere year-round. Sprinkled throughout are a few spotlights on fan favorites, like the begonia; there you'll find the nitty gritty on care and a few fun facts, along with a handful of projects that feature the plant. You'll also see that each plant ingredient listed in the book includes identifying information like the genus (functions like a last name), their more common names (kind of like nicknames), and botanical (scientific) names (similar to first, middle, and last name) to help you find the specific plant pictured.

Don't worry too much, though, if you can't find the exact plant—using something of similar size and shape will create a slightly different but just as lovely arrangement. And once you're comfortable with the theories and design tools presented in this book, you'll be able to customize and create arrangements to suit your every whim. Whether your style is pop art bright or romantic and sweet, the hope is that you'll find the information, inspiration, and delight within these pages to create beautiful living centerpieces.

The Tabletop Garden Toolbox

These are the tools that were used to make the container gardens in this book.

COATED WIRE: To tie soft stems together or to a stake.

DROPPER/SQUEEZE BOTTLE: For watering tiny pots.

GLUE DOTS: These help hold cellophane in place.

MISTER: To water leaves gently.

PAINTBRUSHES/SHAVING BRUSHES: Good for dusting plants.

PLASTIC LINER: To protect surfaces from getting wet.

PRUNING SHEARS OR SCISSORS: Used to trim bigger-stemmed plants.

SCREEN: Covers drainage holes to stop roots from plugging up the hole and preventing drainage.

SKEWERS: To hold up a floppy head of an orchid, or to create a stem for a succulent cutting.

SMALL ANGLED SNIPS: Angled to prune tiny plants in tiny places.

SMALL SNIPS: Used for pruning delicate plants.

SPOON OR TROWEL: Used to transfer soil or gravel from the bag to the pot.

TURKEY BASTER: Allows you to water in tight spaces.

TWEEZERS: Great for reaching into tiny spaces to arrange small plants, sticks, and stones.

WATERING CAN: Designed precisely for watering; easy to fill and easy to pour.

WATERPROOF FLORIST TAPE: Used to seal cellophane or foil to double-protect pots and make them watertight.

WIRE OR TWINE: Used to tie branches together or to tie a stem to a stake.

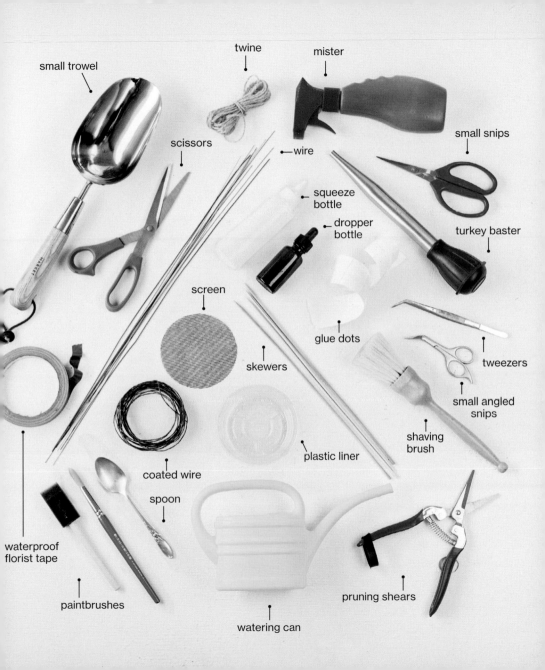

small trowel

twine

mister

scissors

wire

small snips

squeeze bottle

dropper bottle

turkey baster

screen

glue dots

tweezers

skewers

small angled snips

shaving brush

coated wire

plastic liner

spoon

waterproof florist tape

paintbrushes

watering can

pruning shears

Choosing Your Container

Don't be constrained by the notion that plants belong in traditional pots. Look around the house—almost any vessel (and you'll see that that term is used lightly here!) can hold a plant. Bowls, cookware, even cups can accommodate container gardens. Pieces of pipe, picture frames, and branches can all be fashioned so that they house gardens, too.

First, consider the growing conditions. The garden will do best and be easiest to care for when the container and the plant(s) in it match up to some extent. Does the plant like humidity? If so, using a glass jar as an ad hoc enclosed terrarium is a good bet. Does the plant like conditions that are dry? A widemouthed low bowl will do the trick. If the plant craves a pool of water to sit in, make sure the vase is watertight. Consider, too, the plant's growth habit. A twisting, swirling vine loves a trellis to climb—so give it a leg up with a handle to wrap itself around before it trails off in another direction.

Second, consider size. Will the plant(s) fit in the container? Think, too, about the needs of the soil and roots along with those of the plant itself. Like people, most plants can squeeze into something a bit small for a while, but they won't put up with it for very long. That's why there are guidelines for sizes throughout the book.

A basket is a fun alternative to a traditional vase. Just make sure you protect furniture from water leaks with an impermeable liner or dish. Recreate this look on page 50.

Next, consider color. Sometimes the right color combination can make your arrangement sing while a monochromatic one can make it look like single note. Use pattern and accent colors. The bright orange flowers of flaming Katy and Christmas cactus with a similarly hued pattern on the basket (pages 16 and 50) make for a cohesive arrangement.

Finally, think about the overall look. Is your plant design classic or woodsy? A hellebore may look more comfortable in a log vase than in a flashy neon bowl (although sometimes such contrasts look fabulous—and rules are meant to be broken).

Other times it's best to just let the plant(s) show off. The long, lean stems of the orchid, for example, look spectacular when put in a low bowl (see pages 86–87).

WOOD FRAMES are great hosts for plants, whether inside or atop. These are fun to play with, and once plants take root, the frames can even be hung on a wall. They can also be simply set on the table as a cool low, square container.

WOOD BOXES just need to be opened and lined, and away you go with a fun and decidedly unconventional planter.

LOGS AND VASES FROM OTHER NATURAL MATERIALS bring even more of the outside in. Again, these need to be lined to avoid a puddle on the table.

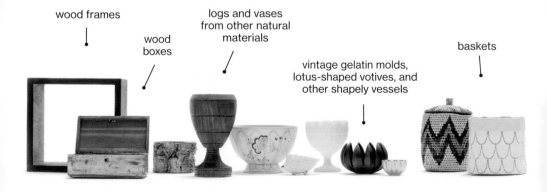

wood frames

wood boxes

logs and vases from other natural materials

vintage gelatin molds, lotus-shaped votives, and other shapely vessels

baskets

VINTAGE GELATIN MOLDS, LOTUS-SHAPED VOTIVES, AND OTHER SHAPELY VESSELS are a chance to have fun and get super creative.

BASKETS feel casual, even a little bit country. Be sure to line them, since they aren't meant to get wet. Some are even flexible and can be stuffed nice and tight with plants.

GLASS TERRARIUMS provide a view from any angle, so be sure the soil and roots look pretty, too. They are best reserved for humidity-loving plants—especially when you keep the lid on.

POTTERY is a classic choice and easy to come by. Most pieces have drainage holes predrilled into them, so be sure to line the inside or set them on top of a plate to protect furniture from moisture.

RUSTIC METAL VESSELS have a charming, weathered, and old-fashioned feel. They can be placed outside, too. Copper and tin, in particular, only look better with time.

PEDESTAL VASES look wonderful with plants that have some drape or droop to them and can add a romantic, even slightly formal, air.

glass terrariums

pottery

rustic metal vessels

pedestal vases

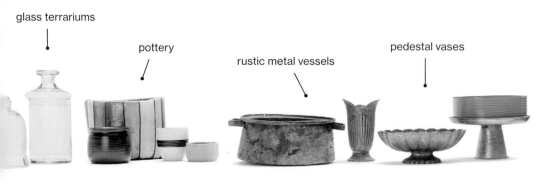

Soil and Amendments

Each container in this book calls for a specific soil type to match the plants and make them happy. In addition to plants and containers, you'll need these things to create your living centerpieces.

SOILS: Soils are mixed with basically the same ingredients but at different proportions, which allows them to hold on to or let go of moisture at different rates. Cactus mix, for example, lets water drain quickly, while potting mix retains water for a bit longer. Some bagged potting mixes contain wetting agents or synthetic materials. Try to steer away from those and stick with organic ingredients. To simplify, the projects in this book call for four types of soil: potting mix, cactus mix, violet mix, and orchid mix.

FERTILIZER: While these projects don't specifically call for fertilizer, it is nice to add some organic fertilizer to your plants. Always check the label for exact amounts and timing and also to make sure it's suitable for your plants.

TOPDRESSING: These pretty additions are a way to top off plant designs and add a bit of polish (they cover the plain old dirt!). The arrangements in this book call for gravel, moss, and bark, but the possibilities are endless, so get creative! Tumbled glass, buttons, and beads, for example, all work beautifully.

Buying the Perfect Plant

Of course plants are available at garden centers—whose reliably knowledgeable staff and large selection will make the trip worthwhile—but they are also increasingly sold at grocery stores, boutiques, pharmacies, and even pet stores. Online nurseries can be a good option for hard-to-find plants. They, too, can provide a wealth of information. Keep these considerations in mind to ensure that the container garden you create will look its best and last as long as possible:

- Each project in this book provides information to help you locate the right plants, including the scientific and common names and plant type (vine, succulent, et cetera). Basic care information, taking into account the specific plant(s) and container, is also provided.

- When in doubt, consult the small plastic tag tucked into the plant's soil to find out how much light the plant wants, how much water it likes, and even when and how often it will bloom.

- Plants can vary greatly in size depending on how and where they are grown. This book calls for a few standard sizes to keep things simple. Some come in round grow pots; some come in square ones. Some will have deep roots and others shallow ones, but if you refer to the size of the grow pot the plant comes in, replicating the effect of each container garden should be a snap. Most of the plants called for in this book are available in 2-inch, 4-inch, or 6-inch grow pots. A few use larger 8-inch and 1-gallon pots, and one even calls for bigger plants (see page 74). There are a few arrangements that use succulent cuttings, which are literally the cut-off pieces of a succulent plant.

- No matter the plant's size, always inspect its foliage (leaves). If the plant is supposed to be green, make sure the leaves are actually green, not yellow or brown. Are they upright? Full of life? Avoid plants with wilting or tearing leaves, as well as ones with notched leaves or nibbled bits, both signs of bugs. Check the leaves carefully on both sides for the black or white specks of insects.

- If you're choosing a plant with flowers, look for one with blooms in various stages. Tight buds and blossoming buds promise a future payoff, and full, open flowers give instant gratification.

- Remember, some plants are seasonal and may be tricky to find at certain times of the year. Though garden centers with greenhouses do allow for a wide selection year-round (some plants are even tricked into blooming out of season), not everything is available everywhere all the time. No matter where you buy plants, some will last for months or even years when taken care of properly, but others, by their nature, will last only a short while.

NOTE: Though some of the plants' common names (as in the case of asparagus fern) make them sound appetizing, the plants in this book are, for the most part, not edible. Some are even toxic.

Plant Care

Because container gardens are composed of plants that will live together in one place, it's best to choose plants that like the same conditions. While some of the designs in this book break that rule, this is a helpful tip to keep in mind. Even if it is a sustainable composition, you might still get an itch to break up and rearrange your containers or decide to move plants from container arrangements into a garden.

When combining different plants in a single container, always consider the soil, the water, and the light. Each of the projects in this book lists that information for the main plant and offers tips on how best to care for that specific container garden. Some plants, like hydrangeas and hellebores, can even be transferred to your garden. Because the information provided here is very general, pair this book with a reliable horticulture reference so you can learn how to care for these plants where you live.

NOTE: When arranging or caring for plants, be aware that many plants, not just those with obvious thorns, can irritate skin. People with sensitive skin should wear gloves, and everyone should wash their hands thoroughly after gardening, whether indoors or out.

Techniques

Follow these general planting principles for all the container gardens in this book unless the instructions specifically state otherwise.

PREPPING THE CONTAINER

Proper drainage is important for a healthy plant. However, you don't want water all over your table. So be sure to take extra precautions and waterproof the pot, vase, or other container accordingly. This will help keep both the arrangement and its setting—be it tabletop, desktop, or wall—in tip-top shape.

Insert a plastic liner into your container, or line it with special waterproofing aluminum and cellophane that are sold at craft stores. Honestly, though, you can use regular aluminum foil and plastic wrap from the grocery store in a pinch. Both are malleable and will conform to the shape of the container. For extra protection, add a strip of waterproof tape to seal it.

Pots with drainage holes need to have those holes covered, especially if the pot will be filled with soil. Simply set a small piece of screen over drainage holes. Not only will this hold the soil in the pot, but it will also assist in drainage by not allowing the roots to block up the holes, so your arrangement can last for months or even years.

As a final waterproofing step, be sure to set pots with drainage holes atop a tray or cork pad. To protect furniture from scratches, place felt dots on the bottom of the tray. Do the same on the bottom of a cork pad, because the plastic layer at the bottom can mark furniture.

Whether planted or just staged (that is, still in their grow pots), plants should sit at the rim of the decorative pot, and sometimes that requires propping them up a bit. Use a small upside-down pot or Bubble Wrap. Cardboard, newspaper, and crumpled-up paper towels are not good choices—they will get wet, lose their shape, and start to rot and smell.

PLANTING ESSENTIALS

MEASURE THE DEPTH

Set the plants next to the pot. Are they the right size? Size them up before you plant so you add enough soil. If you will be staging a pot, measuring the depth will tell you how much prop material you'll need.

USE A FUNNEL OR A SCOOP

If you don't have a trowel, a small flexible cup will help you scoop your soil or gravel into the pot. In a pinch, you can also make your own funnel with a piece of paper and some tape.

MASSAGE THE ROOTS

When you unpot a plant, there may be a tightly bound root system. If so, gently massage the roots to give them space to grow. This technique also works if you need to plant in a pot that is slightly smaller than the original.

PLANT AT THE CROWN

Gently make a small hole in the soil and place the plant inside. Fill in with soil around the base, but don't bury the stem. The soil should hit where the stem ends and meets the roots. Roots don't like to be in the air and stems don't want to live under the soil. Gently tamp down the soil.

The Design Elements

INGREDIENT ROLES

When I create a container garden, I always keep in mind these different design elements. Getting to know and categorizing the elements of your design in this way will help when you want to swap out some plants and are ready to start creating your own designs.

FOCAL SPECIMEN

This is the wow plant, the one to show off, whether it has a huge rosette or a bold color. Let it stand tall, front and center, as with this lady's slipper.

ARCHITECTURAL PLANTS

These provide structure and powerful movement with stern, comparatively tall leaves that
set the frame of the design. The zebra plant
shown here is an austere example.

AIRY ELEMENTS

Fluffy plants provide a little air and give the
design some room to breathe. Plus, they tend
to be eye candy, like this begonia flower.

COLOR

If the focal plant isn't very colorful, a twinkling
of yellow and red (as demonstrated by the
echeveria bloom here) or a dangling tiny trumpet flower adds a bit of pop.

SUPPORTING CAST

Some might call these "filler," but they work just
as hard as the other plants. By adding a range
of texture (and color), plants like the spikemoss
shown here help the focal plant shine.

INGREDIENT TYPES: TEXTURAL

Since so many of the ingredients in a container garden are green, texture is incredibly important. These are the textural elements you'll want to include when creating your own designs.

FANCY FOLIAGE

Some leaves are covered in stripes, swirls, or speckles (as shown on the elephant ear here). Others are crinkly or fold in on themselves.

WISPY

Add movement with plants that swing in the breeze and draw the eye up and out from the center of the container, as with this air plant.

ROSETTES

Not just for roses! Other plants, like succulents and air plants, grow in rosettes with swirling centers that provide focal points. The echeveria shown here is a perfect example.

CLIMBERS

Ivy and other vines (like this ficus) allow the plant to climb out of the container. Up and out is the most common option, but letting these climbers move along a tabletop or around a container is also nice.

DRAPEY

Plants that drape down with ease (like the pictured hoya) are perfect for pedestal planters or hanging baskets. Some will reach down to the floor, creating a truly dramatic effect.

THE GARDENS

Winter Bulbs

PLANTS

Two 6-inch amaryllis
bulbs (*Hippeastrum*) with
unopened and open blooms

Three 4-inch
ripple peperomias
(*Peperomia caperata*)

One 6-inch pink calla lily
(*Zantedeschia*)

CONTAINER AND
MATERIALS

Wood salad bowl, about
16 inches in diameter and
4 inches tall

1 cork protector, 14 inches
in diameter

Plastic bag (a small
garbage bag works well)

4 to 6 clear plastic liners,
one for each grow pot

1. Set the cork on the bottom of the bowl, then line the bowl with a plastic bag. Remove the tallest amaryllis from its grow pot and set it inside a plastic liner in the center back of the bowl.

2. Fill the rest of the bowl with the liners. Set the second amaryllis to the right of the first, then set the peperomia plants along the front of the bowl.

3. Add the calla lily to the left center of the bowl. Fluff the peperomia leaves to conceal the liners. Water each plant about once a week and cut the blooms when they fade.

Ground Cover

PLANTS

One 4-inch ivy
(*Hedera helix*)

One 4-inch dichondra
(*Dichondra repens*
'Emerald Falls')

One 4-inch lobster flower
(*Plectranthus neochilus*)

One 4-inch primrose
(*Primula capitata* ssp.
mooreana)

One 4-inch potato vine
(*Ipomea* 'Bright Ideas
Rusty Red')

One 4-inch Australian violet
(*Viola hederacea*)

Six 4-inch violas: 3 *Viola*
'Frosted Chocolate' and
3 *V.* 'Sorbet Raspberry'

CONTAINER AND
MATERIALS

Vintage metal toolbox,
19½ inches by 8 inches
and 2 inches tall

1 cup potting mix

1. Add a layer of potting mix to the toolbox.

2. Unpot and plant the ivy along the front edge of the toolbox and the dichondra on the front right corner. Drape both plants off the edges. Unpot and plant the lobster flower and the primrose in the back left corner.

3. Unpot and plant the potato vine for contrast in the middle of the toolbox, stretching its vine across the center front to back.

4. Fill in any gaps with the violet and the violas. Let the dichondra and potato vine grow long and prune the primrose to encourage it to rebloom. Water once or twice a week, depending on the temperature of the space (more often in warm climates), keeping the plants slightly moist.

A Forest in Miniature

PLANTS

One 4-inch tongue fern
(*Pyrrosia sheareri*)

One 4-inch mini moth
orchid (miniature
Phalaenopsis)

One 2-inch tropical pitcher
plant (*Nepenthes*)

Two 3-inch air plants
with a pink tint
(*Tillandsia capitata* 'Peach'
and *T. velutina* are nice)

One 2-inch wispy air plant
(*Tillandsia fuchsii* var.
gracilis or *T. filifolia*)

CONTAINER AND
MATERIALS

Shallow pot, 8 inches in
diameter and 3 inches tall

One 1-inch square of
screen

A few lichen-covered
branches

Glue gun

1 cup potting mix

1 cup orchid bark

1. This DIY decorated planter is an integral part of
   the arrangement, evoking a forest floor. Set the
   pot on a table. Cover the drainage hole with the
   screen. Break the lichen-covered branches into
   3-inch pieces. Use the branches to build and
   glue a frame around the pot.

2. Remove the fern from its pot and shake off any
   excess soil. Set it in the center of the pot. Place
   the potted orchid and pitcher plants inside to
   the right and in front of the fern. Fill in any
   gaps with the potting mix.

3. Layer the top with the orchid bark and set the
   air plants on top. Keep moist but not soggy, and
   mist once or twice a week.

Pattern Play

PLANT

One 6-inch begonia with dappled leaves (look for easy-to-find *Begonia bowerae* var. *nigramarga* or *B.* 'Tiger Paws')

CONTAINER AND MATERIALS

Glazed blue flowerpot, sized to match the plant's grow pot

1-inch square of screen

Metal cachepot (decorative holder)

1. Cover the drainage hole in the glazed pot with the screen. Unpot the begonia and replant it in the new container.

2. Place it in the sink and water thoroughly. Let the water drain out the bottom.

3. Set the glazed pot inside the cachepot.

4. Gently tug the leaves so that they flow out and over the edges of the pot. To water, remove the flowerpot from the cachepot, place the pot in the sink, give the plant a thorough watering, let drain, and return. Let the surface soil dry between waterings.

BEGONIA

SOIL: *Potting mix*   WATER: *Keep just moist; allow the surface soil to dry between waterings*
LIGHT: *Bright indirect*

With such a diverse genus, begonias are chosen for their fabulous blowsy blooms or fantastically decorative foliage. Featured here and in the following projects are the fancy-leaf begonias. This bunch has references that include rex, eyelash, escargot, rhizomatous (and so many more). Don't get confused by all the names—just pick one you like and go for it. If their beautiful leaves with swirls of colors and patterns aren't enough to recommend them, fancy-leaf begonias bloom, too. Their airy blossoms are an accent floating above those gorgeous leaves.

Pretty in Pink

PLANTS

One 6-inch asparagus fern (*Asparagus plumosus*)

Two 4-inch begonias: 1 rex begonia in bloom (*Begonia* 'River Nile' is a good choice) and 1 miniature eyelash begonia (*B. bowerae* 'Leprechaun')

One 4-inch coralbells (*Heuchera*); look for a variety with the name 'Lime' in it for this fabulous color

One 4-inch Sprengeri fern (*Asparagus densiflorus* Sprengeri Group)

CONTAINER AND MATERIALS

Decorative bowl, 9 inches in diameter and 6 inches tall

4 to 5 cups potting mix

1. Fill the bowl two-thirds full with potting mix. Unpot the 6-inch asparagus fern and replant it to the left of center at the back of the bowl.

2. Unpot and replant the in-bloom begonia to the right of center in the bowl. Unpot and replant the coralbells to the left of center at the front of the bowl, below the asparagus fern. Unpot and replant the miniature eyelash begonia to the right at the front, below the other begonia.

3. Place the 4-inch Sprengeri fern front and center. Fill the bowl with potting mix to about a half inch below the bowl's rim. Let dry between waterings. Cut back the blooms as they fade, and separate and repot the plants after a few weeks.

A Basket of Begonias

PLANTS

Two 4-inch chocolate cosmos (*Cosmos atrosanguineus* 'Chocamocha')

One 4-inch coralbells (*Heuchera* 'Electra') in bright yellow

Three 6-inch trailing double begonias in pale yellow (try *Begonia boliviensis* 'Bon Bon Sherbet')

One 4-inch begonia with chocolate foliage (*Begonia* 'Black Taffeta' or 'Black Coffee' makes a good choice)

Two 4-inch mirror plants (*Coprosma* 'Karo Red')

CONTAINER AND MATERIALS

Straw basket, 12 inches square and 5 inches deep

Large round glass vase that fits snugly inside the basket

1. Set the potted plants inside the glass vase to assess the arrangement. Start by placing the cosmos and coralbells in the middle. Then surround these with the other plants, working from tallest to shortest as you reach the edge of the vase. Then remove the pots and place them on your worktable in the same design.

2. Unpot and plant each plant, moving from the center outward. Add a layer of potting mix to the bottom of the vase as needed to ensure that the crowns of all the plants are flush with the rim of the vase. Pack the plants in tightly.

3. Carefully place the vase in the basket. Gently arrange any overflowing leaves so that they drape over the basket's edge. Keep slightly moist. As the cosmos fade, the trailing begonias will continue to flower. Once the trailing begonias have finished flowering, plant the chocolate-leaf begonia in a small single pot for a long indoor life. In milder climates the coralbells and mirror plants can go outside, even in winter.

Ingredients shown on next spread

"A Basket of Begonias" ingredients

coralbells

mirror
plant

trailing
double begonia

begonia

chocolate
cosmos

Purple Haze

PLANTS

One 6-inch rex begonia (*Begonia* 'China Curl')

One 4-inch hebe (*Hebe* 'Red Edge')

Two 2-inch miniature peacock orchids (*Pleione*)

Two 2-inch flower dust plants (*Kalanchoe pumila*)

CONTAINER AND MATERIALS

Low pedestal earthenware pot (6 inches square)

¼ cup small lava rock

1 cup potting mix

¼ cup sphagnum moss

1. Pour a 1-inch layer of the lava rock into the pot. Fill the pot two-thirds full with the potting mix. Place the begonia on the left side of the pot; the crown should sit just below the rim.

2. Add in the hebe to the right of the begonia. Create a mound with the orchids and flower dust plants at the front right and center of the arrangement.

3. Fill in any gaps with potting mix and gently press so the plants stay put. Layer the top with the moss. Let dry between waterings roughly once a week. Repot the hebe and the orchids when the blooms fade; keep the flower dust and begonia as is.

An Everlasting Arrangement

PLANTS

One 4-inch creeping fig (*Ficus pumilla*)

Two 4-inch sansevierias (*Sansevieria trifasciata* 'Golden Hahnii' and *S. trifasciata* 'Black Star' are nice choices)

One 2-inch Christmas cactus (*Schlumbergera* 'Christmas Fantasy' has peachy orange blooms)

One 2-inch flaming Katy (*Kalanchoe blossfeldiana*)

CONTAINER AND MATERIALS

Straw basket, 6 inches in diameter

Cellophane

Plastic liner, 6 inches in diameter

6-inch grow pot (with holes)

1 cup cactus mix

1. Line the basket with cellophane and insert the plastic liner.

2. Unpot the creeping fig and loosely replant it in the front area of the 6-inch grow pot (with holes) set on top of the liner in the basket. Drape the vines over the edge. Plant one of the sansevierias in the back, tilting it to the right. Add the second sansevieria in the back of the arrangement. Tip it to the right also so that the two plants fit together seamlessly.

3. Tuck both the Christmas cactus and the flaming Katy into the center and left of center. Add the cactus mix as needed. This is a long-lasting arrangement; remove the planted arrangement from the lined basket to water sparingly.

Natural Neons

One 4-inch blue angel's trumpet (*Eriolarynx australis*)

One 6-inch asparagus fern (*Asparagus plumosus*)

One 6-inch elephant ear (*Caladium bicolor*)

One 4-inch bougainvillea (*Bougainvillea × buttiana* 'Barbara Karst')

One 4-inch Japanese sedge (*Carex* 'Ice Dance')

One 6-inch pineapple lily (*Eucomis comosa*)

CONTAINER AND MATERIALS

Weathered copper cooking pot, about 24 inches in diameter

Plastic liner, such as a small garbage bag

Water-tolerant stuffing, such as Bubble Wrap

1. Set the liner in the cooking pot and fill it with the waterproof stuffing.

2. Leave all the plants in their original pots. Begin with the tallest plant, the blue angel's trumpet, placing it in the center back of the cooking pot as the backdrop and architectural element. Place the asparagus fern slightly in front of the blue angel's trumpet and to its right, allowing the fern to drape over the side of the pot.

3. For a pop of color and fancy foliage, add the elephant ear and bougainvillea to the front center and left of the pot, angling them so that they drape over the edge.

4. Add the Japanese sedge and pineapple lily to the front right of the pot for a striking finish.

5. Water each plant separately, about twice a week, and be sure to keep the elephant ear moist. This arrangement will last about 3 months and can easily be dismantled or rearranged as the plants grow, bloom, or fade.

Ingredients shown on next spread

"Natural Neons" ingredients

bougainvillea

elephant ear

Japanese sedge

blue angel's trumpet

pineapple lily

asparagus fern

English Garden

PLANT

One 6-inch clematis (Regal *Clematis* is shown here)

CONTAINER AND MATERIALS

Stone pedestal planter, 7 inches in diameter and 8 inches tall

One 1-inch square of screen

1 to 4 cups potting mix

One 6-inch square of sheet moss

1. Cover the hole in the planter with the screen.

2. Fill the bottom of the planter with the potting mix. Leave enough room for the plant plus a few inches. Unpot the clematis and set it in the planter. Fill in around it with more potting mix, then add a layer of the sheet moss.

3. Gently untangle the plant from its trellis, if it came with one. Tug lightly at the vines to give them asymmetrical movement.

4. Water the arrangement in a sink until the water runs out from the bottom of the planter. Let it finish dripping. Place it on a waterproof plate if the display surface might become damaged by moisture.

PLANT SPOTLIGHT

CLEMATIS

SOIL: *Potting mix amended with peat moss*   WATER: *Keep just moist to moist*
LIGHT: *Bright direct*

The twisting tendrils and cascading delicate flowers make this vine irresistible. Though it is sometimes sold as a houseplant, it's not meant to live for years indoors. Clematis can be planted outside once the weather warms up in late spring. Keep the roots cool to keep their gorgeous blooms happy.

Blushing Blooms

PLANTS

One 1-gallon dwarf Japanese skimmia (*Skimmia japonica*)

One 6- to 8-inch wood hyacinth (*Ledebouria socialis*)

One 6-inch clematis (Regal *Clematis*)

CONTAINER AND MATERIALS

Faux cement planter, 9 inches in diameter by 11 inches tall

One 1-inch square of screen

1 to 3 cups potting mix

1. Cover the hole in the planter with the screen. Add a few inches of the potting mix and place the unpotted skimmia in the planter. Add more potting mix if its crown sits more than 2 inches below the rim.

2. Repeat the measuring and placing procedure with the wood hyacinth and clematis.

3. After the clematis is planted, tenderly untangle the vine and wrap it around the base of the plants for a compact shape and contemporary look. Water once or twice a week, making sure to let the water run out the bottom.

Classical Asymmetry

PLANTS

One 6-inch hydrangea
(*Hydrangea macrophylla*)

One 6-inch star-of-
Bethlehem (*Ornithogalum
umbellatum*)

One 6-inch clematis (Regal
Clematis)

**CONTAINER AND
MATERIALS**

Wood vase, 11 inches in
diameter and 13 inches tall

Cellophane

Water-tolerant stuffing,
such as Bubble Wrap

1. Set the hydrangea next to the vase to determine how much support will bring the plant's crown to the rim of the vase. Line the vase with cellophane and add the Bubble Wrap. Keep the plants in their original containers but remove any strings, plant supports, or foil.

2. Set the hydrangea to the right of center in front. Tilt it toward the front of the vase, draping the leaves and blooms over the rim. Add the star-of-Bethlehem in the back.

3. Finally, place the clematis on the front left rim, draping its vines down the vase's side. Keep each plant moist, and plant outside once the blooms have faded.

Untamed Jungle

PLANTS

One 6-inch thornless
crown of thorns
(*Euphorbia geroldii*)

Two 6-inch crowns of
thorns with yellow blooms
(try *Euphorbia milii*
'Primrose Yellow')

One 4-inch crown of thorns
(*Euphorbia milii*)

One 4-inch bush sedge
(*Carex solandri*)

Two 4-inch boxwood
honeysuckles (*Lonicera
ligustrina* var. *yunnanensis*
'Baggesen's Gold')

One 4-inch variegated
red apple plant (*Aptenia
cordifolia* 'Variegata')

Two 4-inch echeverias,
green varieties with blooms
(try *Echeveria* 'Ramillete')

CONTAINER AND MATERIALS

Tin bowl, 10 inches in
diameter and 5 inches tall

1 to 3 cups cactus mix

1. Select various euphorbia plants of different heights and sizes for this jungle-like arrangement. Add cactus mix as needed.

2. Start in the center and add the three taller crowns of thorns. Place the large-leafed focal euphorbia in the front. Move to the front and plant the fourth and smallest euphorbia.

3. Taking care to avoid the two thorny euphorbia, plant the softer plants. Start tall again with the sedge and honeysuckles. Plant them at an angle, and weave the honeysuckle stems into the euphorbia stems to create flow in the arrangement.

4. Fill in the bottom layer with the softer variegated red apple plant and the two echeverias. Drape them over the edge to soften the bowl's edge.

5. Keep the arrangement in bright light and water once a week, making sure there is no standing water in the bottom of the bowl.

Ingredients shown on next spread

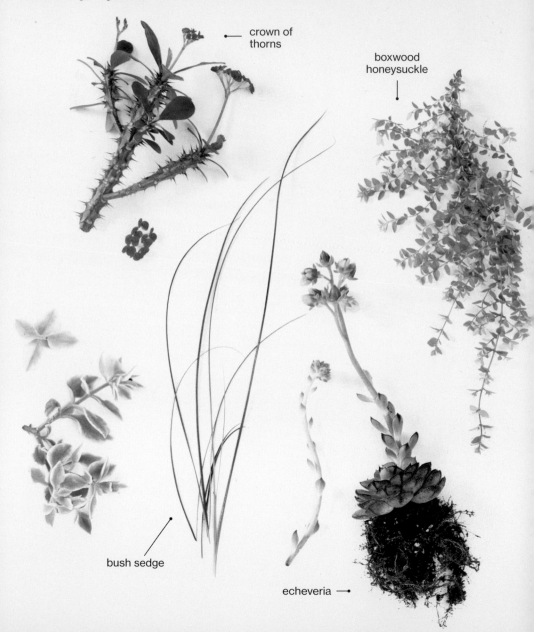

"Untamed Jungle" ingredients

crown of thorns

boxwood honeysuckle

bush sedge

echeveria

variegated red apple plant

thornless crown of thorns

'Primrose Yellow'
crown of thorns

Dancing Ladies

PLANTS

One 4-inch dancing-lady orchid (*Oncidium* Pacific Sunrise 'Hakalau')

One 1-gallon Japanese painted fern (*Athyrium niponicum* var. *pictum*)

One 8-inch xerographica air plant (*Tillandsia xerographica*)

CONTAINER AND MATERIALS

Glazed low bowl, 8½ inches in diameter and 3 inches tall

Plastic liner, 4 inches in diameter

3 cups potting mix

1 cup orchid bark

1. Set the liner into the bowl toward the back, off-center. Fill in with the potting mix around the liner. Place the orchid grow pot in the liner, if it will fit. Otherwise, remove the orchid from the pot and place it in the liner. Fill in around the orchid with the bark, then scoop a small hole in the soil in front and to the left of the orchid.

2. Unpot the fern and set it in the scooped-out hole. Make sure that the plant is level with the rim of the bowl.

3. Cover the soil with bark, and set the air plant on top, in front of and to the right of the orchid. Water each plant once a week, separately.

Tropical Treasure Chest

PLANTS

One 8-inch air plant
(*Tillandsia oerstediana*)

One 6-inch fuchsia
(*Fuchsia*)

One 6-inch elephant ear
(*Caladium bicolor*)

One 6-inch rex begonia
vine (*Cissus discolor*)

CONTAINER AND MATERIALS

Wood box with lid,
14 inches by 10 inches and
5½ inches tall

Plastic bag

5 cups small lava rock

1. Line the box with the plastic bag and pour a thin layer of the lava rock on the bottom. Keep the air plant in its original pot and set it on top of the rocks.

2. Keep the lid of the box open and use it as a backdrop for the plants. Slightly angle the air plant forward so that the inside center is visible.

3. Keep the fuchsia in its container and weave its arching stems into the large leaves of the air plant.

4. Squeeze the potted elephant ear into the box on an opposite diagonal to the air plant. Let the leaves flop over the corner of the box.

5. Finally, insert the potted the rex begonia vine. Eventually this will wind and twine its way through the arrangement.

6. Water each plant separately, giving more moisture to the elephant ear and fuchsia. If one element fades, pull it out and repot it. Make sure the arrangement receives bright but indirect light.

Ingredients shown on next spread

"Tropical Treasure Chest" ingredients

fuchsia

air plant

rex begonia vine

elephant ear

Wooded Whimsy

PLANT

One 6-inch hellebore
(*Helleborus*)

**CONTAINER AND
MATERIALS**

Log vase, 7 inches
in diameter

Plastic liner that fits snugly
in the vase

Two 2- to 3-inch rounds
of cushion moss

1. Insert the liner into the log vase and set the plant (still in its pot) into the vessel. If needed, stage the pot (see page 25) to align the crown with the top of the vase.

2. Cover the soil around the exposed pot with the moss. Water once a week or when dry—keep just moist. When the hellebore is done blooming, plant it outside in the garden. The serrated leaves are amazing, especially when the blooms are long gone.

PLANT SPOTLIGHT

HELLEBORE

SOIL: *Potting mix*   WATER: *Keep just moist*   LIGHT: *Low to bright indirect*

This early bloomer is also called Lenten rose or Christmas rose. Soft, bobbing flower rosettes dangle above its tough, serrated leaves. There are so many species and hybrids of this plant, but one thing for sure is that they are all lovely! The colors of the blooms range from gorgeous mauve to a brighter green. Some have spots and some a double bloom to boot.

Rugged Blooms

PLANTS

Two 3-gallon aeoniums (*Aeonium* 'Cyclops' is a nice choice)

One 1-gallon aeonium (*Aeonium arboreum* 'Atropurpureum' and *A.* 'Blackbeard' make nice choices)

One 2-gallon silver spear (*Astelia chathamica*)

One 1-gallon hellebore (*Helleborus*)

CONTAINER AND MATERIALS

Aluminum bucket, 17 inches in diameter and 18 inches tall

2 bricks or other sturdy objects for staging

Plastic liner that fits snugly into the bucket

One 10-gallon plastic pot with drainage holes (16 inches by 16 inches)

One 12-quart bag potting mix

1. Determine the height of the bucket in relation to the plants. Pick appropriate materials to stage the plants (see page 25) and place them in the bucket. Set the liner on top.

2. Unpot the aeoniums and plant them in the plastic pot, adding the potting mix as needed so that the plants are sitting at the correct height. Then add in the draping silver spear.

3. Place the pot in the decorative container, then unpot and plant the hellebore into it. All these plants can handle shade and might like it best on a shady deck after spending a few weeks inside. Water when dry.

A Study in Contrasts

PLANTS

One 6-inch hellebore
(*Helleborus*)

One 4-inch coralbells
with purple leaves
(*Heuchera* 'Palace Purple'
or similar variety is fine)

One 6-inch East Indian
holly fern (*Arachniodes
simplicior* 'Variegata')

One 4-inch forget-me-not
(*Myosotis scorpioides*)

**CONTAINER AND
MATERIALS**

Vibrant bowl, 10 inches in
diameter

5 to 10 cups potting mix

1. Choose a vibrant color bowl to offset the yellow in the fern, the dark purples of the coralbells leaves, and the mauve accents in the hellebore.

2. Unpot all the plants. Add enough of the potting mix to the bowl to allow for the plants' crowns to sit level with the rim of the bowl.

3. Plant the hellebore first. Loosen the roots and release soil to make room for the other plants.

4. Plant the coralbells at the front and left of center, tilting it at a slight outward angle so the leaves drape over the edge of the bowl.

5. Repeat the planting method with the fern, placing it right of center at the front. Tuck the forget-me-not into the front center of the bowl.

6. Fluff the ferns up and out and stake any hellebore flower stems that need to be firmly rearranged to face forward. Keep the arrangement moist by watering it once or twice a week. The forget-me-not will fade first—after the hellebore blooms fade, disassemble the arrangement and repot the plants separately, indoors or out.

Ingredients shown on next spread

"A Study in Contrasts" ingredients

hellebore

coralbells

forget-me-not

East Indian holly fern

Lush Lantern

PLANTS

One 4-inch lipstick plant (*Aeschynanthus lobbianus* 'Variegata')

One 4-inch mock orange (*Philadelphus*), in bloom

One 4-inch basket vine (*Aeschynanthus* 'Black Pagoda')

One 4-inch rabbit's foot fern (*Humata tyermanii*)

CONTAINER AND MATERIALS

Candle lantern with a door and vents at the top, 6 inches wide by 15 inches tall

Cellophane

1 cup potting mix

1. Add a layer of cellophane and a 1-inch layer of the potting mix along the bottom of the lantern. Unpot the lipstick plant and place it at the back. Gently pull the vine stems up through the top vents and let them drape.

2. Unpot and add in the white-flowering mock orange, placing it in the back left.

3. Finish with the basket vine and the rabbit's foot fern, draping them out of the front of the open lantern. Gently tug the roots of the fern out and let them wrap along the edge of the lantern. Although the other plants would prefer to be only slightly moist, the mock orange likes it moist. Water twice a week. Replant the mock orange when the blooms fade.

Ruffles and Roses

PLANTS

One 6-inch sansevieria (*Sansevieria kirkii* var. *pulchra* 'Coppertone')

One 4-inch orchid (*Oncidium* Twinkle 'Red Fantasy'), about to bloom

One 6-inch begonia (*Begonia* 'Black Coffee' is a nice choice)

One 4-inch miniature rose (*Rosa*)

CONTAINER AND MATERIALS

Large copper bowl, 18 inches in diameter

2 cups potting mix

One 12-inch square of sheet moss

1. Add the potting mix to the bowl. Lay out the plants and unpot all but the orchid.

2. Start with the sansevieria, planting it at an extreme, almost horizontal, angle, on one side of the bowl. You may use the entire plant or just a section of it, as it may split when unpotted.

3. Repeat with the potted orchid on the opposite side, then fill in the back area with the begonia.

4. Snugly plant the rose in between the sansevieria and the orchid so that its full blossoms are front and center. Fill in any gaps with the moss to cover up the soil. Gently tug the begonia leaves so that they intermingle with the other plants.

5. While the rose's blooms will be fleeting, the orchid will bloom and the blossoms will last for months. Let the arrangement dry between waterings about twice a week, giving more moisture to the roses. Avoid soggy roots.

Ingredients shown on next spread

"Ruffles and Roses" ingredients

miniature rose

orchid

begonia

sansevieria

Jewel Tones

PLANTS

One 6-inch masdevallia orchid (*Masdevallia*)

Three 4-inch masdevallia orchids (*Masdevallia*)

CONTAINER AND MATERIALS

One two-tone shallow ceramic bowl, 13 inches in diameter

3 cups orchid mix

3 feet coated wire

One 12-inch square of sheet moss

1. Mound the orchid mix in the center of the bowl, making sure the inside color is visible. Unpot the biggest orchid and place it in the center of the bowl. Gently pat down the orchid so that it will stay put.

2. Fill in with the remaining orchids. Gently tamp the soil to keep them upright.

3. Unroll the coated wire and wrap it around the base of your plants to keep them upright. Wrap a smooth layer of the moss around the base so that it forms a perfect circle. Keep the arrangement moist and cool.

PLANT SPOTLIGHT

MASDEVALLIA ORCHID

SOIL: *Potting mix or orchid mix*   WATER: *Keep just moist*   LIGHT: *Low to bright indirect*

These animated, bat-like flowers have arching necks and dancing heads that dangle atop their skinny stems, making them look like bird beaks as the pods grow. Keep these plants cool and moist. They do not tolerate dry heat.

Long and Lean

PLANTS

Two 4-inch masdevallia orchids (*Masdevallia*)

One 6-inch blue flax lily (*Dianella caerulea*)

Three 4-inch earth stars (*Cryptanthus* 'Black Mystic')

Four 4-inch nerve plants (*Fittonia*)

CONTAINER AND MATERIALS

Scalloped clay pot, 10 inches in diameter and 18 inches tall

One 1-gallon container

Plastic liner, 7 inches in diameter

1. This is a large staged arrangement in which the plants remain in their original pots. Measure the plant heights in comparison to the decorative pot (see page 26). Place the gallon container upside down in the pot, then set the liner on top in order to catch water and keep it from dripping out.

2. Begin staging the plants. Set the tall orchids on top of the liner. Make sure the rims of the pots rest slightly below the rim of the decorative pot.

3. Set the blue flax lily in between and slightly behind the orchids.

4. Add in the earth stars around the center-placed plants, then place the nerve plants in the center front of the arrangement. Let the orchids flow and dangle through and around the arrangement. Keep each plant moist.

Woodland Violets

PLANTS

One 4-inch African violet (*Saintpaulia*)

One 4-inch spike moss with a branching habit (*Selaginella*)

CONTAINER AND MATERIALS

Block vase of California black walnut, with a cavity 6 inches in diameter and 3 inches deep

Plastic liner or melted wax

1 cup violet mix

1. Protect the bottom of the vase with a plastic liner that fits snugly, or melt a candle and spread the wax over the exposed interior to create a flexible yet waterproof layer. Size up the violet with the pot. If necessary, add the violet mix to the liner.

2. Unpot and gently loosen the soil around the violet. Plant it in the plastic liner, leaving half of the container open for the spike moss. Tilt the plant so that its leaves drape over one flat edge of the vase.

3. Repeat the planting process with the spike moss. Keep the arrangement moist.

Fiery Sunset

PLANTS

One 6-inch lily (*Lilium*)

One 6-inch star-of-Bethlehem (*Ornithogalum dubium*)

Two 4-inch baby rubber plants (*Peperomia obtusifolia* 'Variegata')

CONTAINER AND MATERIALS

Colorful bowl, 10 inches in diameter and 6 inches tall

5 cups potting mix

1. Set the plants around the bowl to determine how much of the potting mix you'll need to bring the crowns of the plants flush with the bowl's rim. Unpot and arrange the plants, with the tall plants in the back and the shorter ones in the front.

2. Add the appropriate level of potting mix and plant the lily.

3. Plant the star-of-Bethlehem in front and to the right of the lily. Place the baby rubber plants in front and to the left, tilting and draping them over the rim of the bowl. Keep the arrangement moist. Replant or compost the lily and star-of-Bethlehem once they fade.

Vintage Trough

PLANTS

Two 4-inch English ivies
(*Hedera helix*)

One 2-inch New Zealand
hair sedge (*Carex comans*
'Frosted Curls')

One 4-inch thyme-leafed
fuchsia (*Fuchsia thymifolia*)

One 4-inch blooming
lewisia (*Lewisia cotyledon*)

Three 4-inch red stars
(*Rhodohypoxis* 'Pintado')

CONTAINER AND
MATERIALS

Refurbished vintage
wood gutter

1 cup potting mix

1. Pour a small amount of the potting mix into the container. Unpot and split the ivy plants into sections and plant them along the length of the container. Unpot and plant the upright New Zealand hair sedge toward the right of the container, between two sections of ivy.

2. Unpot and plant the fuchsia to the left of center, between two sections of ivy, allowing its leaves to drape over the front edge.

3. Fill in the remaining spaces with the blooming lewisia and red stars. Water until moist, about once or twice a week, and keep in bright light.

Springtime in a Bowl

PLANTS

Three 4-inch hyacinths
(*Hyacinthus orientalis*)

Two 4-inch tulips (*Tulipa*)

Two 4-inch rosy maidenhair
ferns (*Adiantum hispidulum*)

Four grape hyacinths
(*Muscari*)

One 6-inch fern
(*Polystichum* is easy to find)

**CONTAINER AND
MATERIALS**

Ceramic serving bowl,
8 inches in diameter and
3½ inches tall

5 cups potting mix

One 6-inch square of sheet
moss

1. Add enough of the potting mix to the bowl to allow for the plants' crowns to sit level with the rim of the bowl.

2. Unpot all the plants.

3. Begin with placing the three large hyacinths in the center. Mound the potting mix below them, if necessary, to give them a bit more height. Plant the tulips to the right of the hyacinths so that they lean slightly over the edge of the bowl, creating a sense of movement.

4. Plant the ferns and grape hyacinths in the front and left of the bowl, filling in the spaces. This will add fluff and texture. Pluck small bits of moss and tuck them in between plants to cover exposed soil. Water evenly until moist. After the blooms fade, repot and keep the ferns.

All in a Row

PLANTS

8 tulip (*Tulipa*) bulbs of varying heights, beginning to bloom

CONTAINER AND MATERIALS

Low metal tray, 10 inches long

½ cup of small lava rock

3 cups of potting mix

1 cup of decorative gravel

1. Select tulip bulbs that have begun to sprout but have not yet fully bloomed. Part of the wonder is watching them grow. Pour a thin layer of lava rock into the tray, then add a sprinkling of the potting mix.

2. Unpot the bulbs and carefully separate and plant them in a line across the tray. Add more potting mix if needed to partially cover the bulbs.

3. Scoop on the decorative gravel so that it slightly covers the bulbs and completely covers the potting mix. Keep moist and cool.

PLANT SPOTLIGHT

TULIP

SOIL: *Potting mix or orchid mix*   WATER: *Keep just moist to moist*   LIGHT: *Bright*

The bulbs for these ephemeral bursts of color are easy to find in grocery stores and nurseries in spring and early summer. Their long stems and blossoms evolve and change daily as they reach for the sky, then pop, stretch some more, and finally droop and fade away.

An Early Summer Centerpiece

PLANTS

Two 4-inch coralbells
(*Heuchera* 'Fire Alarm')

Three 6-inch flowering
tuberous begonias
(tuberous *Begonias* are
usually found in the summer
months filled with loads of
blooms)

Three 6-inch tulips (*Tulipa*)

CONTAINER AND MATERIALS

Tin vase, 12 inches in
diameter

Metal cake stand,
12 inches in diameter

5 cups potting mix

Stakes (if needed)

Twine (if needed)

1. Place the vase on top of the metal cake stand.

2. Add the potting mix to the vase until it is three-quarters full. Unpot the plants and set them aside. Begin with the coralbells, placing them at the front edge of the vase on the left, positioning the leaves so that they drape over the rim. Place the blooming begonias next to the coralbells, on the right, letting their rosettes float over the right edge.

3. Mound the back center of the vase with more potting mix. Then fill in the back of the tin with the tulips. Separate the bulbs if necessary to spread out the blooms, but be careful to keep the flowers' roots intact. Gently warm up and massage the tulip stems to make them turn in a desired direction. You may add a hidden stake with twine if more aggressive manipulation is necessary.

4. The tulips will grow, stretch, move, and finally fade. When they do, disassemble the arrangement, compost the tulips, and repot the begonias, which will bloom all summer long.

A Dramatic Duo

PLANTS

One 6-inch multifloral lady's slipper (*Paphiopedilum* Robinianum gx)

One 6-inch rex begonia with dark leaves (*Begonia* 'Helen Teupel' is a nice choice)

CONTAINER AND MATERIALS

Glass box

Block of wood

2 plastic liners

1 cup decorative gravel

Two 12-inch squares of sheet moss

1. Place the wood block in the glass box to prop up the plants. Place the gravel in the bottom of the plastic liners and set them on top of the block.

2. Unpot the lady's slipper and plant it in one of the liners. Place the potted begonia in the other liner.

3. Then tuck the sheet moss in between the walls of the box and the plants to hide the liners. Thoroughly water the lady's slipper once a week, but do not let the roots sit in water. Remove the begonia about twice a week to water it.

Velvet Menagerie

PLANTS

One 6-inch velvet elephant ear (*Kalanchoe beharensis*)

Five 6-inch cyclamens (*Cyclamen persicum*)

Two 6-inch dwarf kangaroo paws (*Anigozanthos*)

One 1-gallon Chinese meadow rue (*Thalictrum ichangense* 'Evening Star')

Two 4-inch panda plants (*Kalanchoe tomentosa*)

Three 2-inch millot kalanchoes (*Kalanchoe millotii*)

One 4-inch hebe (*Hebe* 'Red Edge')

Two 2-inch lavender scallops (*Kalanchoe fedtschenkoi*)

CONTAINER AND MATERIALS

Bronze light fixture flipped over, 14 inches square and 7 inches tall

One piece of screen, big enough to cover the small opening in your vessel

One 8-quart bag cactus mix

1. Cover the small opening with the screen.

2. Add the cactus mix until the container is about two-thirds full, then add more to the center to create a mound.

3. Unpot all the plants. Plant the 6-inch velvet elephant ear in the back, to the right of center. Make sure it rests at a high point of the arrangement.

4. Plant the cyclamens next, placing them around the sides and in front of the elephant ear.

5. Plant the kangaroo paws and rue next, with the kangaroo paws to the left and the rue to the right of the elephant ear, and tucked behind the cyclamens.

6. Fill the front of the container with the panda plants, the millot kalanchoes, and the hebe. Add in the lavender scallops. Fluff the cyclamens and the rue blooms to highlight the plants' pinks and browns. Let dry between waterings about once a week.

Ingredients shown on next spread

"Velvet Menagerie" ingredients

lavender scallop

cyclamen

panda plant

velvet elephant ear

millot kalanchoe

Chinese meadow rue

hebe

dwarf
kangaroo paw

ACKNOWLEDGMENTS

Sophie de Lignerolles's talent shines in so many ways—her work as my right hand in many of these creations was, as always, invaluable. Paige Green's talent behind the camera is astonishing. Her sense of light, angle, and composition made the arrangements in this book come to life. Her gracious personality and lovely spirit made each photo session delightful.

To Janet Hall, a huge thanks for suggesting my name to Kitty Cowles. Kitty took a project I had dreamed about and made it a reality. I can't thank her enough for her introduction, insight, and support. Thanks to Lia Ronnen, Bridget Monroe Itkin, and Elise Ramsbottom for their trust, vision, and patience. With their guidance, this book flourished. And thanks to Suet Chong and Nina Simoneaux for their design creativity, and to Carson Lombardi, Keonaona Peterson, and Sibylle Kazeroid for their keen eyes on the copy. Thanks to Molly Watson for her magic with words. She has a beautiful knack for keeping my voice while making me sound so much better.

Writing a book and running a business was only possible with the help of Lila B. Design's Max Schroder, Cliff Fogle, Shannon Lynn, Brandon Pruett, Mimi Arnold, and Sarah Green who kept our design studio rolling and the creativity flowing.

My mom, dad, and sister nurtured me with their kindness. Many thanks to Tom Ortenzi, who has volunteered his time through PCV to help me run my business. To Lawrence Lee, Robin Stockwell at Succulent Gardens, and SF Foliage for their openness to my plethora of plant questions. Thanks to everyone at that beautiful and eco-friendly estate where I worked in my early days of gardening for influencing me and my style in such positive ways.

To Stable Cafe in San Francisco, a million thanks for graciously hosting our photo shoots in the lovely cafe's courtyard and letting us turn the back garden into a plant design studio. A piece of this garden is in every photo. Finally, thanks to the friends, artists, and businesses who provided a few of the noteworthy containers for the projects in this book: Heath (pages 87 and 96), Pseudo Studios (page 91), Old & Board (page 95), and Joe Chambers (page 104).

INDEX

Adiantum hispidulum, 97
aeonium, 74
Aeschynanthus lobbianus, 80
African violet, 90
air plant
    Dancing Ladies, 66
    A Forest in Miniature, 38
    Tropical Treasure Chest, 69
amaryllis, 34
Anigozanthos, 105
Aptenia cordifolia, 63
Arachniodes, 76
Asparagus densiflorus, 42
asparagus fern
    Natural Neons, 53
    Pretty in Pink, 42
Astella chathamica, 74
Athyrium niponicum, 66

basket vine, 80
begonia (*Begonia*), 41
    A Basket of Begonias, 45
    A Dramatic Duo, 102
    An Early Summer
      Centerpiece, 101
    Pattern Play, 41
    Pretty in Pink, 42
    Purple Haze, 48

    Ruffles and Roses, 83
begonia vine, rex (*Cissus
    discolor*), 69
blue angel's trumpet, 53
blue flax lily, 89
bougainvillea, 53
boxwood honeysuckle, 63
bush sedge, *see* sedge
buying plants, 21–22

Caladium bicolor, see
    elephant ear
calla lily, 34
Carex, see sedge
caring for plants, 23
Chinese meadow rue, 105
Christmas cactus, 50
Cissus discolor, 69
clematis, 56
    Blushing Blooms, 59
    Classical Asymmetry, 60
    English Garden, 56
containers
    choosing, 17–19
    prepping, 24–25
Coprosma, 45
coralbells
    A Basket of Begonias, 45

    An Early Summer
      Centerpiece, 101
    Pretty in Pink, 42
    A Study in Contrasts, 76
cosmos, 45
creeping fig, 50
crown of thorns, 63
Cryptanthus, 89
cyclamen, 105

dancing-lady orchid,
    see orchid (*Oncidium*)
design elements, 28–31
Dianella caerulea, 89
dichondra, 37

earth star, 89
East Indian holly fern, 76
echeveria, 63
elephant ear (*Caladium
    bicolor*)
    Natural Neons, 53
    Tropical Treasure Chest,
      69
elephant ear, velvet, *see
    Kalanchoe*
English ivy, *see* ivy
Eriolarynx australis, 53

Eucomis comosa, 53
Euphorbia, 63

fern
        asparagus, *see* asparagus
        fern
        East Indian holly, 76
        Japanese painted, 66
        Polystichum, 97
        rabbit's foot, 80
        rosy maidenhair, 97
        Sprengeri, 42
        tongue, 38
Ficus pumilla, 51
fig, creeping, 51
Fittonia, 89
flaming Katy, *see Kalanchoe*
flower dust plant,
    see Kalanchoe
forget-me-not, 76
fuchsia
        Tropical Treasure Chest,
        69
        Vintage Trough, 94

grape hyacinth, 97

hebe
        Purple Haze, 48
        Velvet Menagerie, 105
Hedera helix, see ivy
hellebore, 73
        Rugged Blooms, 74
        A Study in Contrasts,
        76
        Wooded Whimsy, 73
Heuchera, see coralbells
Hippeastrum, 34
honeysuckle, boxwood, 63
Humata tyermanii, 80
hyacinth, 97

hyacinth, grape, 97
hydrangea, 60

Ipomea, 37
ivy
        Ground Cover, 37
        Vintage Trough, 94

Japanese painted fern, 66
Japanese sedge, *see* sedge
Japanese skimmia, 59

Kalanchoe
        An Everlasting
        Arrangement, 50
        Purple Haze, 48
        Velvet Menagerie, 105
kangaroo paws, dwarf, 105

lady's slipper, 102
lavender scallops, *see*
        Kalanchoe
Ledebouria socialis, 59
lewisia, 94
Lillium, 93
lily, 93
lipstick plant, 80
lobster flower, 37
Lonicera ligustrina, 63

maidenhair fern, rosy, 97
Masdevallia, see orchid,
        masdevallia
mirror plant, 45
mock orange, 80
moth orchid, 38
Muscari, 97
Myosotis scorpioides, 76

Nepenthes, 38
nerve plant, 89

New Zealand hair sedge,
        see sedge

orchid (*Oncidium*)
        Dancing Ladies, 66
        Ruffles and Roses, 83
orchid, masdevallia, 86
        Jewel Tones, 86
        Long and Lean, 89
orchid, moth, 38
orchid, peacock, 48
Ornithogalum, see star-of-
        Bethlehem

panda plant, *see Kalanchoe*
Paphiopedilum, 102
peacock orchid, 48
peperomia, ripple, 34
Peperomia obtusifolia (baby
        rubber plant), 93
Phalaenopsis, 38
Philadelphus, 80
pineapple lily, 53
pitcher plant, 38
planting essentials, 26–27
Plectranthus neochilus, 37
Pleione, 48
Polystichum, 97
potato vine, 37
primrose, 37
Pyrrosia sheareri, 38

rabbit's foot fern, 80
red apple plant, 63
red stars, 94
rex begonia vine, 69
Rhodohypoxis, 94
rose, miniature, 83
rosy maidenhair fern, 97
rubber plant, 93

Saintpaulia, 90
sansevieria
    An Everlasting
        Arrangement, 50
    Ruffles and Roses, 83
Schlumbergera, 50
sedge
    Natural Neons, 53
    Untamed Jungle, 63
    Vintage Trough, 94
Selaginella, 90
silver spear, 74
skimmia, Japanese, 59
soil and amendments, 20
spike moss, 90
Sprengeri fern, 42
star-of-Bethlehem
    Classical Asymmetry, 60
    Fiery Sunset, 93

tabletop gardens, 11–12
    buying plants for, 21–22
    caring for plants in, 23
    container choices for,
        17–19
    container preparation for,
        24–25
    ingredient roles in, 28–29
    ingredient types in, 30–31
    planting, 26–27
    soil and amendments
        for, 20
    toolbox for arranging, 14
techniques, 24–27
textures, 30–31
Thalictrum ichangense, 105
thyme-leafed fuchsia, *see*
    fuchsia
Tillandsia, see air plant

tongue fern, 38
tools, 14
tuberous begonia, *see* begonia
tulip, 98
    All in a Row, 98
    An Early Summer
        Centerpiece, 101
    Springtime in a Bowl, 97

violas, 37
violet, African, 90
violet, Australian, 37

wood hyacinth, 59

Zantedeschia, 34